About the Author

In 1961, high school literacy class, the teacher asked his students to write a short story. The teacher liked Fawaz's story very much and asked him to read it to the class. His teacher and students loved it. After high-school, Fawaz had to choose between literacy and less money or engineering and more money. He entered the Faculty of engineering, the world of constructions and then the world of business. Immigrated to Canada in 1996 with a wonderful family, a wife and three children. Several years later he went to live in Beirut a known city for business and literacy.

Dedication

Among all sciences and majors I have always been fond of the study of Civil Engineer. In Civil Engineering universities we learn how to be closer to "Logic" and "Reality" and learn how to use the Science of Math to answer questions and problems in the realistic life. Our professors used to emphasize on teaching us what is called the "Engineering Sense", which is the sense that alerts you when you deviate from the logical results and makes you settle and base your answers only on numbers and logic. In this occasion, I would like to advice university students in all different majors to try to be close to their teachers even outside the classes and lectures to listen to their advices and guidance. I never forget those circles where we used to gather around the professor after the lecture to ask and discuss theories and opinions, whereas in those circles students get to hear the genius ideas that teachers generously pass to the next generation, those great ideas which conclude their practical experiences and instruct them to look and analyze deep in the science and not only take it as is.

I dedicate my trial in this book to my teachers, maybe to their blessed spirits…

Fawaz Al-Maasarani

THE SPEED OF TIME

AUSTIN MACAULEY
PUBLISHERS LTD.

A CIP catalogue record for this title is available from the British Library.

ISBN 9781784551698

www.austinmacauley.com

First Published (2015)
Austin Macauley Publishers Ltd.
25 Canada Square
Canary Wharf
London
E14 5LB

Printed and bound in Great Britain

Acknowledgments

I thought they are only mere ideas that I used to write down on a paper to relive myself from them coming back to me and occupying my thoughts, those ideas that I could escape from in the day but keep me up at night. A deep happiness generates inside me whenever I open those papers to read what I have written and start adding more until it became my passion.

May, my beloved wife, who used to peak at my writings, decided that those ideas are very precious and should be well saved and protected than just stay on papers so she started to type it on her laptop. She used to think that those ideas must be published, and publishing them is more a duty than it is a right.

Preface

When the creator decided to push the pointer of time to run nonstop, it was a natural result that the big bang took place and so the movement started...and after it the expansion...and the development...and life...it was all a matter that is infinitely small in size but with a very massive energy extending to infinity...

When time stops electrons sticks to the nucleus of the atom causing matter to disappear...causing the universe to disappear! When time moves again, electrons rotate around the nucleus of atom and give the atom its size and dimensions...give the universe its size and dimensions...the dimension of the universe extends from zero to a very far distances and maybe to infinity.

Humanity has always focused on measuring the past and the future, but rarely someone tried to measure the present. Is the present a day, an hour or a mere second...does it even exist?

We measure the speed of everything relative to time, but what is the speed of time? Einstein talked about the speed of light travelling in space among stars, but he never touched upon how that light travels simultaneously from the past towards the future at the same speed.

THE SPEED OF TIME introduces these topics and many ideas through email discussions that go back and forth between Aziz Kamal and Ahmad Abdellatif, two friends that met only in the digital world and lived a "spiritual relationship".

This book is an invitation for thinking and contemplation, this is the reason why there is a blank page at the end of each chapter; to invite the reader to write down what ideas each chapter has given her. Therefore, I kindly ask you to send us what you write about each subject to the following page www.facebook.com/thespeedoftimebook

Where we can all study and discuss the proposed ideas on this site. I hope you don't belittle your ideas for they might be a step forward to understand the world around us.

Table of Contents

Chapter 0: Introduction by Ahmad Abedelatif

This book actually presents and is based on the ideas of my friend Aziz, who I have never met, a feat of the age of electronic meetings, which put forth intellectual communication first and sometimes first and last.

I have chosen this collection of emails from about four hundred correspondences I exchanged with Aziz during a period of almost two and a half years. Considering the variety of subjects that we tackled, I have chosen specific subjects that serve the general frame work of this small, rich book.

This collection of emails has caused some kind of inconsistency of thoughts. Therefore, I hope you forgive me for that, in spite that I have done my best to organize the correspondences according to the subjects and chapters that I have chosen because they were the most stimulating for ideas and evoking for discussions.

Chapter 1: The Present is a Time that has no Dimension

From: Aziz Kamal.

To: Ahmadabedlatif@gmail.com

Cc:

Sub: The Present is a Time that has no Dimension.

Dear Ahmad,

When you draw, you need to draw lots of lines, and these lines whether straight or curved have different thickness according to the need and function to produce your beautiful paintings. By the way, I would like to mention to you that I have seen a lot of your paintings and I have written many comments about them which I will send you someday, which will be yet another rich subject to discuss.

I am really not a painter, but I have good taste in art. I feel that the lines, their indications and colors penetrate deep into my senses in a mystical and touching way. It takes me days in order to understand them in my thoughts and feelings so I can express them in words. All that makes me feel I am able to communicate with you in a beautiful, mystical and in a profound way all at the same time. So here I am again, as usual, drifting away from the subject I started to talk about that is the lines you draw have different thicknesses, bold lines and fine lines.

Nevertheless, sometimes you draw borderlines such as the borderline between two colours and this borderline performs a very important function but at the same time it has no dimension, which means it has a width that equals zero. It is in fact a very important line that really exists and divides the green and the blue colours and without it you cannot accomplish your painting and express your ideas and feelings, but still it remains a borderline with a width that equals zero.

Now imagine you draw for me a painting, half black on the left and half white on the right. Let us say that the white area represents the future and the black area represents the past; thus the borderline that has no dimension is the present which means the present is the time that has no dimension and is in other words, the border between two colours or two times and its weight equals zero.

Every word, gesture or glance is directly transmitted from the future to the past. Thus, any human being who tries to capture a minute in the present, absolutely fails to do so simply because it does not exist. For any word, gesture or glance is transmitted directly from the future to the past, so the present is the border between two colours, two extremely broad times. The past spreads endlessly behind us and the future extends endlessly ahead of us, therefore we realize that the present does not exist…it has no dimension…it is simply a zero which is a borderline whose task is to transmit the moment from the future to the past. It is just like a scanner, a time reader that reads what is written on the time page and transforms it from the future to the past and from something anonymous to something known.

How important and vital is that? Indeed it is very important because that is "life" itself! The borderline that has no dimensions and that reads the anonymous and makes it known and that keeps nibbling the future to transform it to past is the life we are living.

Therefore, I say here that the present is a time frame with zero dimensions and that we live a very big number of zeros. For the past consists of an endless number of "Present Zeros" as well as the future, but at the end they are merely zeros.

Nevertheless, the science of mathematics has a different point of view, which we should pay attention to here, because life and mathematics are intermingled and life without figures loses specifications and credibility. It is like saying that the year consists of many number of days without specifying, or saying that the sun is far far away.

So what does mathematicians say about all that? They say that when you multiply $0 \times \infty =$ it is not a zero!

I always thought that: $0 \times \infty =$ zero, because any large number: Large number $\times 0 = 0$.

Yet in mathematics (∞) is not part of the real known simple or compound numbers therefore we can't multiply it in a normal way in any of the digital groups.

"Infinity" cannot be considered a number no matter how big it is. Infinity is a concept outside the framework of mathematics, which we deal with on earth and within our limited standards. In other words, as I understand it, infinity is a "dynamic state" and not a fixed number.

I really don't want to get caught in the mathematical mazes and mathematical proofs, but I had to tackle on this field in order not to reach the result that:

Life = Zero

as is a little better because the result raises life up to the indeterminate form, where mathematicians say that:

$\infty \times 0 =$ Indeterminate form.

and the indeterminate form means indeterminable.

Allow me now to leave alone the mathematical concepts and try to express myself in a more realistic way by saying that life is close to an undetermined form, a form that is lost between zero and infinity, and the "Present", our reality, is a time that equals zero. So, the "Past" is a time that came and passed and the "Future" is a time that will come and might not, and finally the "Present" which is the reality that we live and feel, is a time that has no dimension, which why we feel that life is an illusion that cannot be grasped.

From: Ahmad Abed Latif

To: Azizkamal@yahoo.com

Cc:

Sub: The Present is a Time that has no Dimension.

Dear Aziz,

I couldn't comprehend yet the impact, was it a fact or a theory, of the concept that you are aiming: that the present equals zero, but I feel when you say that you dissolve life, just as if you are dissolving a spoon of salt in the ocean. It simply dissolves without leaving a trace. You are turning life into a great group of zeros, but at the end they are mere zeros. I actually live life and see it as a combination of happiness, tragedies, feelings and events which fill me with vitality, strength, weakness, and contradictions including love, hate, admiration, humiliation and reverence.

These are the colours which I enjoy, and see in nature and employ in my paintings; therefore, I will not allow you to sum up all that into a group of zeros which will turn my life into a null, absent, non-existent life.

I see that you are taking the role of a physicist, who analyses matter and studies it. I really pity scientists, and the sciences, which always analyse and study everything.

Today I saw a beautiful rose, a charming rose indeed, and weather that was in the past or in the future, and in spite of its short span of life, the rose exists in all three times past, present and future. In any form and any time, this rose is glorying its creator and signalling signs and gesturers about the beauty of this universe and its massive dimensions that is out of human scope of understanding. Those dimensions exist in the soul "Spiritual Dimensions". If you only sit down on your laze chair to enjoy a musical piece composed by Chopin, it will make you feel life in its all dimensions, ancient dimensions, modern ones and the charming ones. All that is beyond the reach of your analysis about time and measuring the present!

From: Aziz Kamal

To: Ahmadabedlatif@gmail.com

Cc:

Sub: The Present is a Time that has no Dimension.

Dear Ahmad,

I read with passion your interesting and beautiful email. I don't want to praise your ideas for you have already expressed them with words from your heart. Nevertheless, that doesn't stop me from saying again that in spite of the wonderful times that we live in among beauty, the beauty of creation and our magnificent feelings, this still does not prevent us from sometimes feeling the absurdity of life and that life is merely a great illusion in which we live.

Here I would like to concentrate on two points, which constantly and insistently impose themselves upon us:

1- The dimension of the "Present" that transforms the future to past, equal zero. From this zero our wonderful life, with its vast span extending to the infinite past and infinite future, was created.

2- That obscure and profound feeling which overtakes us when we think of life… the feeling of the illusion of life…. needs a logical explanation.

Why do we feel life is an illusion?

The explanation is that: "the present has no dimension".

The present equals zero and life is a big group of zeros and that is the secret of the feeling that life is an illusion. You really don't need to be highly educated or a sensitive artist in order to feel that life is an illusion. Any simple person can feel this by merely being alone by himself and thinking of his past and the events that he encountered in his life. Thus, he feels that the most important events in his life have become points in memory.

Having said that, let us observe how people try to hold on to the moment and stop time by taking pictures. You see them trying to record a happy moment so it can stay happy, where one can go back to it whenever he or she wants.

This is very similar to how leaders in ancient eras, turned to portrait painting trying to immortalize themselves.

From: Ahmad Abed Latif

To: Azizkamal@yahoo.com

Cc:

Sub: The Present is a Time that has no Dimension

Dear Aziz,

The period of the present time might be very small, but it exists, which explains the accomplishments in which we live. An example of these accomplishments is the civil law, which is an accumulation of knowledge throughout human life, and another example is the pyramids, which were built and kept as a historical witnesses.

There are endless examples of accomplishments which all prove that life is more than a group of zeros.

From: Aziz Kamal

To: Ahmadabedlatif@gmail.com

Cc:

Sub: The Present is a time that has no Dimension.

Dear Ahmad,

Here you are raising the subject of work and production, as if you are saying: "How can these zeros of the present time produce such a civilization?" I hope you don't think that I am denying "life", for it exists and extends back many long years into the past, and it will extend many long years into the future, while the present is a scanner which moves at the speed of time to transmit the future to the past.

The present is a scanner, which has no dimension, and is the borderline between two times which are endlessly extended in both directions. While work is every thought, building or any form of life which continues from a certain point in the past to a point in the future. This might continue and go on for minutes, for hours or for years and thus it will produce and build. Despite the fact that present time is a time that has no dimension, no one can deny its existence and importance.

For the "past" with all its glories, memories, inventions, wars and arts was once present; and the future with its dreams, hopes, and brightness will also pass through the point of the present. It is the present, which will transmit the future into the past at great speed. We have to discover this speed, which moves us through time and takes us to the endlessness of time or the infinity of time.

This present, which has no dimension, is similar to the zero in math. For the zero also has no dimension but it is a very important number, which separates the negative and positive numbers, and it is the largest of all the negatives numbers that extend to infinity.

Chapter 2: The Speed of Time

From: Aziz Kamal

To: Ahmadabedlatif@gmail.com

Cc:

Sub: The Speed of Time.

Dear Ahmad,

No doubt time moves at a certain speed, the borderline between the past and the future move along to nibble the future time and to transform it to past at great speed, but what is this speed?

Here at this point, it comes to mind, that if I start thinking that we rotate around the earth and that the speed of time could be the speed of the earth's rotation around itself every 24 hours, that is:

The perimeter of earth ÷ 24 hours, that is:

40000 ÷ 24 hours = km 1500/ hour.

Perhaps we should add to this speed, the speed of the rotation of the earth around the sun, and the speed of the solar course in the galaxy then the speed of our galaxy, the Milky Way, in space, and the speed that is the result of the expansion of the universe. However, all these speeds are insufficient earth measurements.

There is no speed for time because the speed of things are measured by two elements which are the distance and time according to the equation:

The speed= distance/ time

For time is one of the two elements of measuring speed, thus how can time have a speed in the earthly concept? Nevertheless, time moves, and everything that moves has a speed, therefore how can we measure its speed?

All speeds are relative speeds, which are measured according to time. There is only one sign that I know that refers to the speed of time which was mentioned in Einstein's theory of relativity, which says that time stops for those who reach the speed of light. This is the only scientific reality that we can rely on because whoever moves at the speed of light, time stops for him. That means that the relative speed between the speed of light and the speed of time= zero. That is, light and time move at one speed that also means that they are moving towards the same direction.

Time moves towards the future, that is, from the past to the future, through the present. Moreover, light also moves towards the future, for there is nothing that can move towards the past. One must stop at this point to give the subject full thought and consideration, for we understand that light departs from a certain star to move through space, but since we are discussing the topic of the speed of time, therefore we must add to this that light departs at a certain moment to move towards the future and it arrives at a moment that is after the one it departed from. This means that light moved from the past towards the future and it cannot be otherwise, that is, it cannot move towards the past.

Therefore, one must be careful that the earth speeds that we calculate are relative speeds; if we say, for example, that we are moving at the speed of 100km/ per hour. That is, we travel through 100 km per hour of time, but this time is a moving element, which moves at a certain speed where the speed equations become only relative earth equations. At the same time, we would have moved from the past towards the future. Thus we say here, that light moves towards the future as well as time.

The Speed of Time Equals in a certain way the Speed of Light

But there is no earthly measurement for time speed, because the speed of light is measured by: distance\ time, and time speed loses its position since it is not an earthly speed. It is the most important earth factor that we live each moment, but for ambiguity, it doesn't have a measurement to determine it and give it its human meaning. For how and where can we find a measurement and a precise meaning for this speed?

The astronomers say, suppose there were a spaceship moving towards a black hole at an accelerated speed, and that this space- ship had a light that flashed once every second, and we observe it with a very big telescope to see these flashes. Every time the spaceship approaches the black hole, we will see that these flashes will expand to once every two seconds, then every three, then every four, then every hour. Thus, until we no longer see any flash, and when we no longer see any flash, that means that the spaceship is moving outside or inside the black hole at a speed that equals the speed of light.

Here we can say, that the great gravity of the black hole attracts everything. It does not allow even light to escape from its gravity, and this is another type of movement of the spaceship at the speed of light.

There are two interpretations for this, or rather two sides to it: The great gravity...

And the great speed (the speed of light) and here we can realize (partly) the relation between matter and gravity with time. Here I would like to explain my own opinion as the following:

Some scientists say that the great gravity of the black hole does not allow light to escape from it, and thus we cannot see anything except black in these holes; rather we see the particles around it then they disappear when they reach it.

The explanation for this phenomenon is that we can see the object before it enters into the black hole scope because the speed of this accelerated object is much less than the speed of light. Any object moving at a speed less than the speed of light is visible for one to see, while any object travelling at the speed of light is invisible and cannot be seen; thus this means that the objects in the black holes move at the speed of light.

Nevertheless, what happens to the objects that move at the speed of light and which must be in the black holes? When a certain object travels at the speed of light, time stops in this object because the speed of light equals the speed of time, and when the time stops the electrons inside the atom stop rotating because they rotate according to a certain speed and this speed becomes zero.

Speed=Distance/Time so when time becomes zero the distance travelled also becomes zero and the speed zero and that means that the centrifugal force (the centrifugal force = Constant X Mass X Speed) which fixes the electrons to a certain distance from the nucleus inside the atom equals zero and the gravity force remains the only effective strength in the atom which makes the electrons stick to the nucleus. As a result, the object becomes small with great density and therefore the volume of the object moving at the speed of light tends to littleness and its density becomes infinite because its weight stays as it is, unless it is transformed to energy. The atom goes towards zero because the nucleus consists of smaller atoms and this is a movable state in zero, as well as the movable state in the expansion of the universe or universes. Therefore, the density in the black hole increases and thus its gravity increases.

Thus, matter returns to zero and it becomes easier to recreate it and propel it to a certain speed in order to make it go back to life by the power of the creator.

We go back and say that when time stops we would have reached the origin of matter in the black hole, and considering that matter is one of the aspects of energy or one of the types of energy, we would have reached the origin of energy. Here, we can imagine or explain that matter is born and dies there in these mystical black holes, which are black holes that resemble mirrors that produce matter on one hand and the opposite of matter on the other hand.

There, the process of creating matter is performed by the conversion of non-matter, and energy is created from the opposite of energy. In time zero, because it moves an object at the speed of light, time has no measurement for its speed but it has an indicator, which is the speed of light.

Thus, according to the theories of some astronomers, galaxies are created in black holes, then after a while, they return to the black holes so they can die in them. This happens in extremely lengthy and separate times according to our measurements, but these separate and long times are a set of zeros: $\infty \times 0 = 0$ and here exists creation.

For creation is to create something from zero, and considering the set of times are zero, and the substance and the non-substance equal zero, the Creator alone knows how to perform that.

From: Aziz Kamal

To: Ahmadabedlatif@gmail.com

Cc:

Sub: The Speed of Time.

Dear Ahmad,

If the speed of light equals in a certain way the speed of time, that explains why the speed of light is the absolute speed that one cannot exceed.

The theory of relativity says that the speed of light in space is the maximum speed, that one cannot exceed it. Scientists explain this by a couple of examples, such as:

Let's suppose that a car is driving at a speed of 100 km/ per hour and a person shoots a bullet in the direction of the moving car, taking into consideration that the speed of the bullet leaving the gun is 1000 km/ per hour and then the same person turns around and shoots another bullet in the opposite direction of the moving car. If we measure the speed of the bullet in the first case you will find that it equals1100km / per hour, while in the second case you will find that the speed of the bullet is 900 km / per hour and that is due to the fact that the speed of the car is added to the speed of the bullet in the first case and is subtracted from it in the second case.

This is the logical sequence of the subject, but what would happen if the gun were replaced by a light source? The theory of relativity says that the situation would be different, for the speed of light remains stable in both cases and doesn't change whatever the speed of the car.

Einstein considers the constant speed of light in space one of the basic laws in the universe.

Here we can confirm that the speed of light is the maximum speed because the speed of light equals in a certain way the speed of time. If someone were able to move at a speed that exceeds the speed of light, that would mean that he moves from the present to the future at a speed that exceeds the speed of time; that is, he will move towards the future faster than time. That is, he will see and find out what will happen in the future.

This is, of course, impossible…. you can't surpass the laws of life: the future is unknown until it becomes the present. The speed of present time towards the future equals in a certain way the speed of light, and the speed of light cannot be surpassed, and the future is unknown until the line of the present surpasses it to become the past.

Thus we say: The speed of time is the ultimate speed and the speed of light is the ultimate speed and these two speeds are somehow equivalent!

Chapter 3: The "Zero" and the "Present"

From: Aziz Kamal

To: Ahmadabedlatif@gmail.com

Cc:

Sub: The "Zero" and the "Present"

Dear Ahmad,

When we study time, we find that the zero in math is the alternative and the equivalent to the present in the scope of the study of time and what applies to the zero mathematically applies to present time in life. Therefore, the zero deserves from us more contemplation and study.

There are two concepts about the zero. The first is simple when we say:

1-1=0

While the other is more complex when we say that the zero is a point between: $-1/\infty$ and $+1/\infty$

Thus we begin.

The zero is not null but it rather exists and is substantial and essential, for it is the borderline that divides positive and negative numbers.

It is the zero, which is infinitely small, that is to say, whenever we find a smaller number, which is closer to zero, there would be a number that is even closer to zero. Therefore we see that the zero is a "moving state", which is not exactly static it is like ∞, Every time there is a big number there was a bigger number and thus even closer to the ∞ that is to say that ∞ is a "moving state" and thus we become more convinced that there is nothing constant in life and everything moves towards certain infinity.

Here we can mention that mathematicians consider the zero a miracle for it is an important element of the elements of interpreting the mysteries of the universe. The zero has no multiples and no parts and if you multiply it or divide it by a number it results in a zero, but if we divide a number/by zero we get infinity, that is a non-existent number in the field of natural and complex numbers.

All the modern computer sciences depend on 1 and zero for the zero exists at a value that equals zero, and it is the beginning of the positive numbers as well as the end of the negative numbers. Here it is exactly like the present, the beginning of the future, and the end of the past, and it exists and has no dimension and therefore one cannot feel it or grasp it. It is like an illusion that passes us; it creates everything and yet is merely nothing.

Here we should pay attention to that, most of the time; there is a specific result for unlimited numbers. For instance, I can walk a distance that consists of an infinite number of an infinitely small distances, which is:

$\infty \times E$ (infinitely small) = not determined

But when I walk 100 m I will have:

$\infty \times E$ (infinitely small) = 100m

I should mention that in the study of zero, when we observe a time or an object coming from $+\infty$ to $-\infty$, from the extreme far away future to the faraway past, it must pass by zero or the "present".

$-\infty \longleftarrow \text{------- zero -------} \longrightarrow +\infty$

Also, we find that before that moving object passes zero, it reaches very close to zero, but it is not a zero for it comes from $-\infty$ to -billion meter then to -1 meter then it starts with fraction numbers and passes at a distance of $-1/10$ meter, then to $-1/$ billion meter, then to $- 1/\infty$ of a meter. It then reaches zero and begins with the positive numbers with $+1/\infty$ of meter, and thus we see that zero is a number located between two numbers that are not real and which present a non-fixed moving state exactly as ∞ but the zero remains a turning point from the negative to the positive. At the same time this is a very important changing from the negative to the positive and the opposite happens in an infinitely small number.

There is a border and not a line that divides between the incoming from the -∞ to +∞ and this border is of course the zero.

For the incoming number from -∞ passes by zero where the number moves from the infinitely small number before zero to the zero then to an infinitely small number after zero, then this infinitely small positive number gets bigger until it reaches +∞ that is infinitely big.

The object, which is moving from zero towards − ∞ and the object that is moving from zero towards +∞ will definitely meet. For these two defined objects launched towards opposite directions must meet in a certain place and a certain time and they must meet in a remote zero as they were launched from a zero in front of us. Because straight lines do not exist, but are actually curved lines, this means that the infinities must meet.

As well as the infinitely small and the infinitely big numbers, when we explore the inside of matter we find that there are smaller and smaller parts, and every time we invent a bigger telescope we can see the universe has bigger extensions and the ends of the bigger and the smaller ends must meet in a certain place or a certain time or in a certain form.

This nature in which the parts of the atom or the parts of the nucleus of atoms meet "is the energy"

It is well known that splitting the atom releases great energy, and the expansion in the universe needs great energy, and this energy is a small part of the invisible effect. It is part of God's power by which this universe was created. The infinitely big meets the infinitely small.

The negative infinity -∞ meets the positive infinity +∞ and this is very significant.

The intersection point of infinities must be zero and that means:

That zero is the beginning,

And zero is the end,

And that God created this universe from zero,

And will return it to zero,

Every human being can feel that he came from zero,

And is moving towards zero,

Like every human being…. like everything….

Zero is the secret of the universe,

As well as the "present time" is the secret of life,

From here we should wonder about the relationship between matter and time.

Chapter 4: The Relationship between Matter and Time

From: Aziz Kamal

To: Ahmadabedlatif@gmail.com

Cc:

Sub: The Relationship between Matter and Time.

Dear Ahmad,

Based on the assumption that the speed of objects inside black holes reach the speed of light and time stops, and that the stars and planets are formed and died in those black holes, makes us wonder what is the relationship between time pausing and the creation of matter or its destruction. The relationship is that matter consists of atoms, and atoms consist of nucleus and electrons, which rotate around the atom, and from this rotation matter gets its dimensions. For when time stops the rotation stops because the rotation happens at a certain speed. The electrons rotate in a certain speed, and this speed is the distance / time and without time or with time = zero the equation becomes number / zero= ∞ or indetermination. When the electrons in the atoms of matter stop rotating, the centrifugal force is expelled and gravity remains, for the equivalence between the centrifugal force and the gravity ends, and at this point the electrons stick to the nucleus and matter becomes very dense and very small. If we imagine that the nucleus of the atom also consists of smaller atoms and similar to the atoms that we know, at this point matter becomes even denser and with a smaller volume, and this is an infinite moving state. Whereas every atom inside it consists of smaller atoms, and as the universe expands towards largeness, matter gets smaller towards zero.

Here we understand the relationship between time and matter for whereas there is no time there is no matter, or matter that the creator can form however he wants. There is consensus among scientists that the "Big-Bang" is the beginning of the universe, however none of them touched upon the reason behind this "Big-Bang".

The Big-Bang happened when the creator first pushed the clock hand. The time that we are living, the past, the present, the future and the Big-Bang itself is a natural result of the beginning of "Time". Before the start of "Time" electrons were slicked to the core of the atom, and for electrons to rotate they need speed, and there is no speed without time, so the time generates movement and movement gives atom its dimensions…gives the universe its dimensions!

Therefore, without time there would be no matter, without time matter contracts to infinity small, without time the universe contract to zero, which was the case before the start of time!

From that we discover that time is the fourth dimension of matter!

There is no matter without the dimension X,

There is no matter without the dimension Y,

There is no matter without the dimension Z,

There is no matter without the dimension Time!

The relationship between matter and energy remains, and is there a form of matter in this transformation? That is, when matter moves towards zero, it discharges energy exactly as what occurs when the atom is split and atomic energy is released. And thus, when time stops, matter is transformed to energy, and whenever we move away from time zero the energy is transformed to matter, and whenever we approached the time zero at the speed of light, matter is transformed to energy.

Is that true, and is the huge quantity of energy is the spirit?????

Chapter 5: Life was created from Zero

From: Aziz Kamal

To: Ahmadabedlatif@gmail.com

Cc:

Sub: Life was created from Zero.

Dear Ahmad,

Life has a tendency toward creation, enlargement and expansion, the expansion of the universe…

It does not have a tendency toward termination, death and contraction, for death is not an element of termination and ending, it is rather a developing and an improving factor. On the basis of that, death is a stage of transition of the living to other fields where the body is no longer needed. The body has consumed its life and lost its validity. The spirit is energy, and energy does not have the ability of termination, and the body is matter and matter also does not have the ability of termination but it is transformed to other non-living matter. Thus the body is transformed to dust... returns back to dust.

We cannot imagine another life that we can move to like paradise or hell. Let us free ourselves from our earthly scientific logic. Let us free ourselves from the knowledge that we have... Let us also free ourselves from being proud that we are practical scientific people. Let us get out of life and look at it from a distance to see how strange our life is, like if I'm there outside the life trying to convince my neighbour that life consists of the intermixing of energy and matter "Soul and body".

I spent days describing this to my neighbour until he thought I was hallucinating, for why do we exclude the existence of another life that includes all creatures who were created throughout history.

Life has a tendency to creation, enlargement and expansion

Creation is a constant permanent movement.

Creation did not occur only once.

Creation occurs every day, even every moment.

Spirit is energy.

Energy is the source of everything.

Creation... everyone admits and acknowledges creation.

Creation means to produce something from zero, while on the other hand, transforming, amending and modification is not creation. Only creation makes something from zero, and here we go back and say that the universe was found and exists in a zero digit, nothing but zero.

From zero, the universe was found. And from zero everything we know was created.

From zero, the universe was found… and from zero everything we do not know was created.

Energy and matter+ anti energy and matter= zero

This leads us back to the line of time.

The borderline that divides between the past and the future is called the present. It is a line that has no dimension and its thickness equals zero.

The line that is a border and is not a line, a border that has no dimension…

And this is the miracle of creation.

Here we go back and say that life is time and without time there is no matter.

Time is a group of moments whose number is infinite.

But in this book we discover that the dimension of all these moments is zero.

And $\infty \times$ zero = zero

In this zero, God breathed life as energy.

And with the beginning of time, energy was transformed into matter.

And from energy and matter we were granted this wonderful life.

Chapter 6: The Future is written

From: Aziz Kamal

To: Ahmadabedlatif@gmail.com

Cc:

Sub: The Future is written.

Dear Ahmad,

Leave your house and walk in the streets and look into peoples' eyes, observe all their moves, look profoundly to see that each one is moving according to certain given information that he cannot deviate from. Each one receives the information from his five senses and from his brain and acts according to the information as if he were programmed. He carries out a role that was designed for him, and cannot deviate a hairsbreadth from it.

Thinking about this subject seems excluded, for we see behaviour is random, but if we look more profoundly at someone's movements and actions, you feel he is under control and moves according to what is written.

From: Aziz Kamal

To: Ahmadabedlatif@gmail.com

Cc:

Sub: The Future is written.

Dear Ahmad,

When the borderline between the past and the future, which we call the present, passes a certain moment, then this borderline which has no dimension, exterminates the future and makes it the past and thus ends it and turns it to a dead memory.

The energy for this moment dies and moves to another form or place.

And the matter remains.

This borderline severely attacks the future moment-by-moment transforming it, along with its wide far away horizon, to a point in memory.

It transforms the future, which is full of hope, expectation, ambition, planning and motivations.

The future is sometimes full of bright colours and, at other times, dark colours, but in all cases it is the future that includes life with all its sweetness and bitterness that makes the heart jump with expectation and hope. It changes the future with a blow that reminds me of an airplane breaking the sound barrier and making this terrifying sound although there is no barrier to break between this entire clamour and roar. In this way, the borderline between the future and past transforms the moment from a future moment to a past moment silently and smoothly, but in great violence that will kill and erase all bright and dark colours of time without distinction to make it a point in memory, a memory of a person on its personal level and the memory of nations on the level of nations and the memory of history on the level of humanity and the memory of the universe.

Is it a great shock that occurs every moment to the extent that we have gotten used to it and no longer notice it, although it slowly kills us, or by an extraordinary speed moving us to our doom. Within this killing shock, there is the secret of life and the secret of death, which move along the borderline between the future and the past at a great speed we could not and we cannot understand its falsehood or its value.

Now suppose we have a very large super computer and we were able to load it with a vast amount of information which would include all data in one certain moment of time; data that includes everything about the movement of each galaxy in space to the movement of each atom in a neglected piece of iron, to the cry of a newly born infant. In this moment since we would have all the information that took place in the universe, without doubt our computer would give us what will happen in the next moment, and, depending on what will happen in the next moment, it could tell us very precisely what will happen the moment after it.

The principle is simple and clear...

Every event in a certain moment is a definite result of the data, which exist in the moment before it.

These are the endless chains of events that happen in every moment in a definite precise way as a result of the data of the moment before it; that is, events occur as a result of the data of the moment before them.

Thus, we say that whoever started creation and pushed it towards time had determined from the beginning exactly what would happen as a result of this first push in the following moments which extend until the end of time.

For every moment following this push and for this moment in which this push took place, and every following moment, is the result of what had happened in the moment before it and so on...

So if there was a computer out there, and in fact there is, it would be capable of determining what will happen in the second moment, then what will happen in the third moment, then what will happen in the fourth moment, then what will happen in the endless (infinite) moments. For I am here not questioning that the future of the universe is predetermined, and that what I shall do tomorrow is written and determined...I am proving, unquestionably, that everything is known and determined beforehand and no one can deviate a hairsbreadth from what is written and determined for him, for you are a result, and another result will occur after you, and you are a loop that will be followed by another loop of a chain.

There are people among us who are proud of their attitudes, inventions and wisdom and I tell them that you are merely a result of what has preceded you.

Your brain conducts you and you did not choose your brain, the events and circumstances around you are imposed upon you. Your body's situation is imposed upon you. Everything is imposed upon you. You are a result...

Your behaviour, be it insignificant or important, is imposed upon you...

You are written on the wall of time...

Like everything else.

Everything is written on the wall of time...

The boundary line, which is present...

The boundary line which has no dimension and which separates between the past and the future is the one that scans and reads the future to make it the past.

Here I feel that the urgent question is: "Who am I. What am I...?"

I am the result of the combination of the chromosomes of my mother and father or something like that from the heredity and the result was a child who had a certain specific heritages and descriptions that you cannot deviate from a hairsbreadth.

Then I was raised in an environment where I learned and acquired additions and simple adjustments to my inherited characteristics and thus my character was formed. I could say then that I learned, studied and worked hard to become an engineer.

But, indeed, I behaved according to my character and inherited genes which were imposed upon me and which were transformed to chemical formulations in the brain and electric current nerves which move in the brain according to a precise system which cannot be changed unless the genetic or the acquired data which is imposed on my behaviour is changed in a precise way that cannot be deviated from.

Therefore Sir, stop being so arrogant and self-conceited and admit that you are a "result" and an "episode" of what is written.

I might believe that I could say this word or I could not say it, but the truth is that I cannot but do what I'm dictated to by what happened in the previous moment of complicated effects of many elements such as the situation of my spirit and body as well as another person's reaction and many other outside effects. But every effect can be analysed and restored to its elements and thus the result is definite because the causes are definite too.

They say that whoever wants to understand more about what is happening on our earth, should go out into space and observe it.

Whoever caught within a subject or a case fails to see it in a comprehensive way and will drown into its details and forget the essence of the subject.

Looking at something from a distance has the advantage of the strategic, conclusive view of something. Thus, we learn more from outer space and see more clearly and understand better what is happening on earth.

Whoever studies the universe gets impressed by the precise movement of the planets, stars, solar systems and galaxies, and observes that everything moves according to a great precise system and that there is no place for a lack of purpose or a lack of organization. In this way, we can understand that we move, work, think and talk on earth according to a similar system that has no place for any possibility of change or emergency amendments as a result of ideas that are different than the general context which is predesigned and prewritten.

The planets move according to systems and rules that cannot be changed and they cannot deviate from the courses, which were precisely set for them. Likewise, mankind, animals, earthquakes, volcanoes and the activities of the wind, rain and clouds and everything else, move according to regulations and rules which cannot be changed or exceeded.

Here we go back to say that on the basis of this theory, what is happening at this moment wherever it is, is the definite result of the causes that were found in the previous moment.

Also, what happened in the previous moment is a definite result of the causes that were found in the moment that was before it.

And thus…

Whoever pushed the universe at the first moment is the one who determined what will happen successively moment by moment, without the ability to change, and that is the reason and the result since the first moment until eternity.

God has created this universe and pushed it towards a certain direction. He created the first reasons and the results of these causes are known and written and are still going on.

You cannot change anything, and thus we go back and say that it is possible to write what happened in the past, from the large comprehensive natural history and from the partial human history. Likewise, it would be possible, if we had the ability to know or to read the future, we would be able to write the future. The line without dimension (which has the width of zero) is the scanner or the reader that scans and separates the past from the future.

Chapter7: The Spiritual Relationship

From: Aziz Kamal

To: Ahmadabedlatif@gmail.com

Cc:

Sub: The Spiritual Relationship.

Dear Ahmad,

I'm sorry to tell you that my husband Aziz died two days ago. He died while he was writing you an email and it is clear that the text of this email was not completed.

Since I am certain that no one but you can complete the text, I wanted to tell you this bad news and to send you the text, and perhaps you can finish it as you see fit and return it to me.

I always used to read your emails and admire your ideas and I believe they should be published someday. As Aziz used to tell me, he sent you his ideas and discussed them with you because he wanted people to share and discuss these ideas, and you are the only eligible person to publish these letters someday. I understand that you are an artist and a sensitive musician, which was what my husband liked most in you. He used tell me sometimes that he was in need for someone who could feel and touch his thoughts, that is, for someone who not only understands but for someone who can touch the depth of the idea, and he used to say that he feels that you were able to do so.

I know that you both never met in spite of your friendship through emails for two years and a half due to the distance between you. However, more important than all this is that he felt a kind of inspiration to talk to a person to whom ideas, and communication were more important than looks, age and accent.

To him you were a spirit to talk to and discuss without personal questions. He felt that you shared this high level of communication with him.

I attached herewith the text of the last **Email**.

From: Ahmad Abed Latif

To: Tarakamal@yahoo.com

Cc:

Sub: The Spiritual Relationship

Dear Tara,

I felt deep sadness overcome me to lose the friend whom I always considered a soul mate. I actually never met him and never felt the need for that. On the contrary, I used to feel that meeting would spoil this kind of new human relationship, which the latest technology has granted us.

This technology allows us to talk to a human being and communicate with him intellectually without feeling the need to see him in person and to even feel that you know him and that you can dive into his ideas and feelings and nature more than if you were sitting with him, for here there is no need for pretence, arrogance or acting. You don't feel embarrassed to absolutely be yourself with someone who is hundreds of miles away from you, and shares with you only his ideas and feelings.

Nevertheless, I always had in my imagination a picture of this wonderful person with whom I have communicated for years. Please don't send me a picture of your husband for I want to keep the picture I have of him in my imagination.

I believe, like you, that no one can complete the text you sent me better than I can, and this is the most appropriate conclusion that is closest to his way of thinking.

From: Tara Kamal.

To: Ahmadabedlatif@gmail.com

Cc:

Sub: The Spiritual Relationship.

Dear Ahmad,

Thank you for sending the text that you wrote, which I actually consider very suitable as a completion to Aziz's text. But allow me to tell you that I can't ignore my feminine nature and hide my curiosity and eagerness to see the picture that you drew of my dear husband from your imagination. Therefore, allow me to ask you to send me the picture you have drawn of my husband from your imagination, especially since I know you are an artist, so I can compare it with his actual picture. In this way I will be able to see the impression my husband's ideas left on a sensitive artist like you. It will allow me to see the impression my husband's ideas, feelings and values left on the imagination of a person who knew him more than anyone else.

From: Tara Kamal

To: Ahmadabedlatif@gmail.com

Cc:

Sub: The Spiritual Relationship

Dear Ahmad,

When I received the picture that you drew of my husband, I felt surprised, or even shocked at its closeness to reality. Therefore, allow me to send you his photograph and I beg you to forgive me because I momentarily suspected that you actually knew him at one time.

Now, I hope you will allow me to assume my husband's role and style of emails and try at least this once to contemplate what my husband called the spiritual relationship and psychic communication between people; these mystical forces which are inherent in man and which include hypnosis. All these forces are in harmony with the nature of my husband's letters.

I will restrict my interjected remarks to trying to study the nature of these forces. I have read that scientists have been able to take a picture of the electromagnetic waves that transmit a kind of communication and effect and hypnotic state. We should point out a very important fact that these electromagnetic waves are the scientific explanation of communication and telepathy, as well as of hypnosis, and at the same time they are the explanation of the transmission of light through the void in the vast spaces between the stars where there is no matter. It is also the way that sound is transmitted between the broadcasting station and the receiving set and the picture between the television transmission station and the television set, etc.; and it is all due to electromagnetic waves which are transmitted at the same speed, which is the speed of light. So did these waves transmit Aziz's picture to you as a result of the telypathy between both of you? Are there different kinds of electromagnetic waves, as I believe? Is the time that we live in, the present time, moving from the past to the future at a speed that equals in a certain way the speed of light and is it also being trnsmitted by electromagnetic waves?

Believe me, my great gratitude towards this wonderful man, at least as I consider my husband, encourages me to urge you to publish your correspondence with him, or at least some of it. In this case, I would be very thankful to you for spreading his ideas, which he always considered to be inspiring for others so that they can add to them, discuss them and develop them. He often said he wanted them to reach a great number of thinkers, scientists and all those who love to think and contemplate.

Chapter 8: The Last Email

The infinities meet as well as the paradoxes. For we cannot see or experience the meeting of the infinities. But we can see and experience every day the meeting of paradoxes that we encounter.

One of the most important paradoxes is:

Life is based on injustice.

Wherever you look around you, you witness injustice.

In the news bulletins you see people being persecuted by all kinds of oppression and even being put to death although innocent and without having a hand in what is happening, and without knowing why they are being oppressed. History is full of vile events and unjustified wars.

Nero oppressed Rome for a long period of time and then burned it without asking his people about their sufferings.

Hitler led Germany and Europe and the whole world into wars and tragedies. He wanted to impose his own extreme opinions and was able to persuade the enthusiastic youth, who paid later, along with their counterparts in other nations, gravely for this extremism.

The strong defeats the weak...

The strong eats the weak...

The wolf eats the sheep...

This is the mode of life...

But let us imagine a life without injustice.

All animals including man eat plants like elephants and like vegetarians.

No devouring and no killing.

Fish live on seaweed and other sea plants.

I am not tackling the subject of the balance of nature here, neither I am talking about the great number of rabbits that will fill the earth, for the creator will find a solution for all that, but the life that we want without injustice or killing is not a life!

For without injustice there is no challenge... there is no struggle for survival... and for success there is no motivation and no competition for the best.

Life is based on injustice.

Because, injustice gives life its meaning!

I do not believe you want a life without meaning, without feelings, without fear, without defiance.

We feel injustice when we are part of the situation, but when we go far, far off, into outer space and we look at the scene of life as a whole, at that point we see justice being born from the womb of injustice, and life emerging from the cruelty of death.

For without oppression there is no justice.

Without ugliness there is no beauty.

Without weakness there is no power.

Without stupidity there is no intelligence

Without death there is no life.

Without time moving at the speed of light and us racing it towards our destiny, whether we like it to or not, there is no death over there and there is no life.

As we have already said. Basically without time there is no matter.

Time is the cause of our suffering.

Time is the cause of our existence.

E2 4/18